Nelson International Science
Workbook 4
Anthony Russell

OXFORD UNIVERSITY PRESS

Great Clarendon Street, Oxford, OX2 6DP, United Kingdom

Oxford University Press is a department of the University of Oxford.
It furthers the University's objective of excellence in research, scholarship,
and education by publishing worldwide. Oxford is a registered trade mark of
Oxford University Press in the UK and in certain other countries

Text © Anthony Russell 2012
Original illustrations © Oxford University Press 2014

The moral rights of the authors have been asserted

First published by Nelson Thornes Ltd in 2012
This edition published by Oxford University Press in 2014

All rights reserved. No part of this publication may be reproduced,
stored in a retrieval system, or transmitted, in any form or by any
means, without the prior permission in writing of Oxford University
Press, or as expressly permitted by law, by licence or under terms
agreed with the appropriate reprographics rights organization.
Enquiries concerning reproduction outside the scope of the above
should be sent to the Rights Department, Oxford University Press, at
the address above.

You must not circulate this work in any other form and you must
impose this same condition on any acquirer

British Library Cataloguing in Publication Data
Data available

978-1-4085-1729-1

10

Printed in India

Acknowledgements

Cover: Andy Peters
Illustrations: Maurizio de Angelis, Tony Forbes, Simon Rumble and Wearset Ltd
Page make-up: Wearset Ltd, Boldon, Tyne and Wear

p.20: (ant) andrey Pavlov/iStockphoto, (bird) Worakit Sirijinda/iStockphoto, (dog) Eric Isselée/Fotolia, (earthworm) john shepherd/iStockphoto, (fish) The Dragon/Fotolia, (frog) TessarTheTegu/iStockphoto, (housefly) arlindo71/iStockphoto, (human) Glenda Powers/Fotolia, (snail) Ursula Alter/iStockphoto; p.22: (crab) Kaiya_Rose/Fotolia, (spider) Colette/Fotolia.

Although we have made every effort to trace and contact all
copyright holders before publication this has not been possible in all
cases. If notified, the publisher will rectify any errors or omissions at
the earliest opportunity.

Links to third party websites are provided by Oxford in good faith
and for information only. Oxford disclaims any responsibility for
the materials contained in any third party website referenced in
this work.

Contents

1 Humans and animals — 3

Skeletons — 3
- Activity 1 — 4
- Activity A — 5

Drugs as medicines — 6
- Activity 2 — 6
- Activity 3 — 9
- Activity 4 — 11
- Activity B — 12

2 Living things in their environment — 14

Animals in their habitats — 14
- Activity 1 — 14
- Activity C — 19

Identification keys — 20
- Activity 2 — 20

Human activity and the environment — 23
- Activity 3 — 23
- Activity 4 — 25
- Activity 5 — 27
- Activity D — 28

3 States of matter — 29

Solids, liquids and gases — 29
- Activity 1 — 30

Heating and cooling materials — 31
- Activity 2 — 31
- Activity E — 33
- Activity 3 — 35

Steam — 38
- Activity 4 — 38
- Activity F — 40

4 Sound — 42

Sound making and measuring — 42
- Activity 1 — 42
- Activity 2 — 43
- Activity 3 — 45
- Activity 4 — 46
- Activity 5 — 47

Sound travelling — 50
- Activity 6 — 50
- Activity 7 — 52
- Activity 8 — 53

Materials preventing sound transmission — 54
- Activity 9 — 54

Pitch and loudness — 56
- Activity G — 56
- Activity H — 58
- Activity 10 — 59
- Activity 11 — 61
- Activity 12 — 63

Musical instruments — 65
- Activity 13 — 65

Contents

5 Electricity and magnetism — 66

Constructing circuits — 66
    Activity 1 — 66

Broken circuits — 69
    Activity 2 — 69
    Activity 3 — 70
    Activity 4 — 71

Magnets attract and repel — 72
    Activity 5 — 72

Metals and magnets — 74
    Activity 6 — 74
    Activity I — 77
    Activity J — 78

Introduction

Nelson International Science Workbook 4 provides a complete copy of the *Student Book* activities for all learners to work through.

The activities are marked with 📖 showing the corresponding page number in the *Student Book*.

In addition to the *Student Book* activities, there are extra activities marked, for example, Activity A, that can be done in the classroom or as homework at home. They support the knowledge and understanding gained in the *Student Book* activities.

Chapter 1: Humans and animals

Skeletons

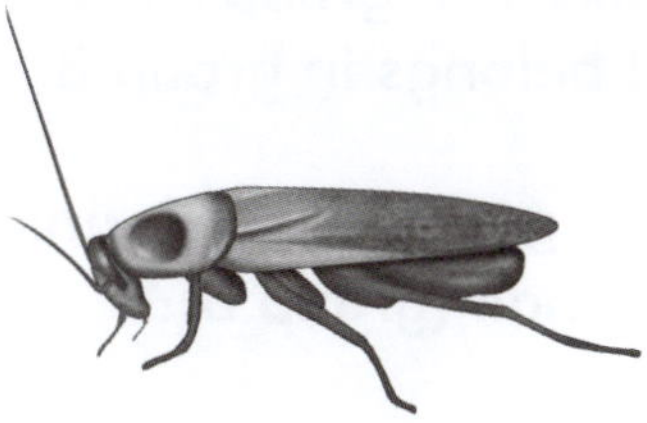

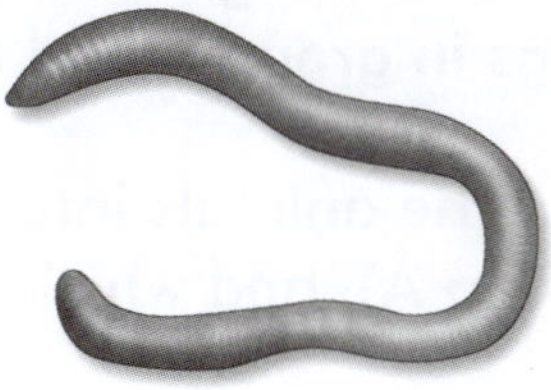

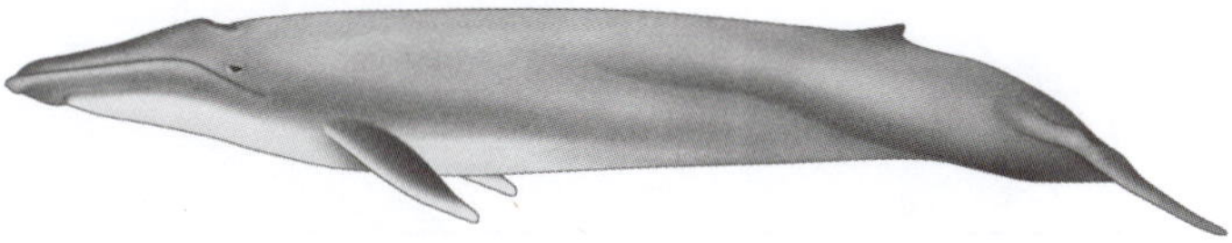

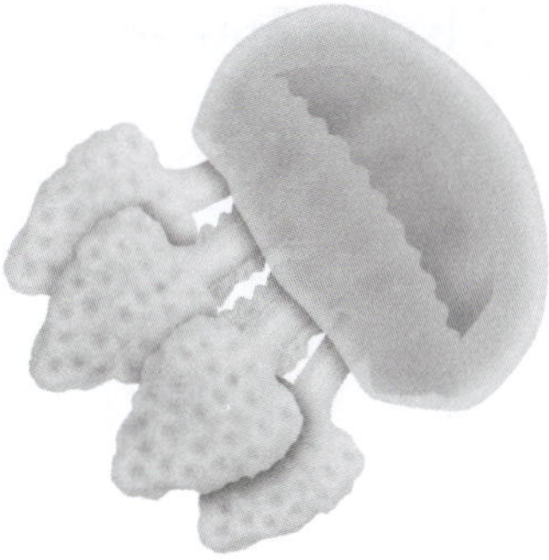

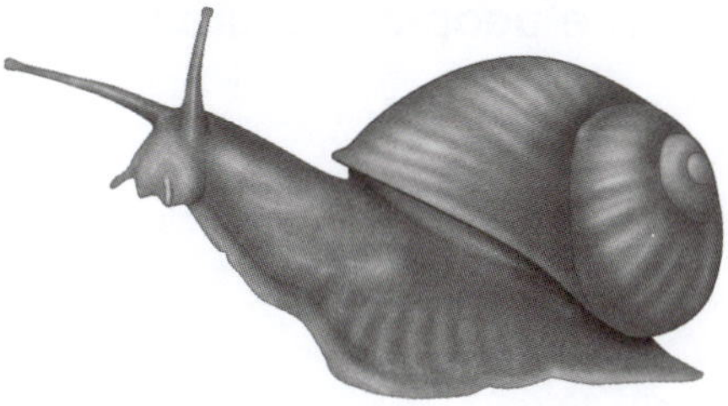

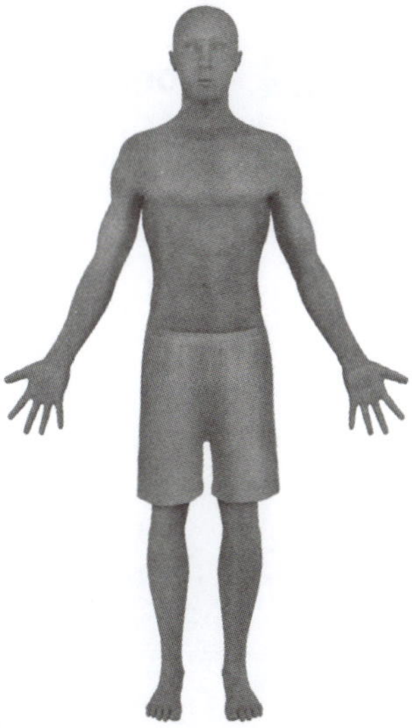

Chapter 1: Humans and animals

Activity 1

You will need: a pen or pencil.

1 Look at the animal pictures on page 3. You can make two groups. For example, the cow belongs in group A and the snail belongs in group B.

2 Try to sort out the rest of the animals into these two groups. Which animals belong with the cow (group A), and which with the snail (group B)?

Write them down.

Group A	**Group B**
_____	_____
_____	_____
_____	_____
_____	_____
_____	_____

3 Show your groups to the people you are working with. Discuss how you have sorted the animals.

4 Share the groups with the class.

Skeletons

Activity A

You will need: a pen or pencil.

Complete these sentences using these words (you may need to use some of them more than once):

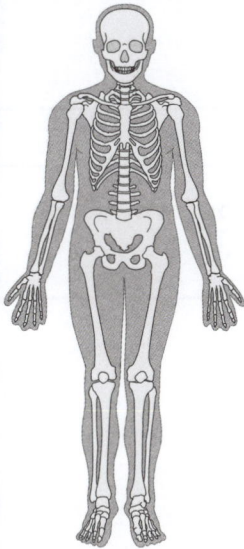

Without a skeleton a human being could not move, breathe or eat

> bend   snails   bones   protection
> external   move   cockroaches
> pair   cows   tendons   lungs
> earthworms   joints   birds   organs
> pull   internal   skull   jellyfish
> muscles   heart   no

1 Two animals that have an _____ skeleton are _____ and _____.

2 Some animals, such as _____ and _____ have ____ skeleton.

3 We have an _____ skeleton, made of _____. Other animals, such as _____ and _____ have skeletons like ours.

4 The bones have _____ which allow us to _____ our bodies. The _____ are attached to the bones by _____.

5 Bones are also important as _____ for important _____ in our bodies.

6 The brain is inside the _____, while the ribs help to keep our _____ and _____ safe from damage.

7 Muscles _____ on bones and this makes them _____. The joints each have a _____ of muscles, one to _____ the joint and one to straighten it.

Drugs as medicines

There are three groups of drugs:
- Prescription drugs
- Over-the-counter drugs
- Prohibited drugs.

Activity 2

You will need: a pen or pencil.

1. Discuss the three groups of drug types. Share your ideas about why some drugs are in one group and not in another. Record them here:

2. Write down examples of each group of drugs.

   _____

   _____

   _____

3. Share your lists and ideas with the class. Add to your lists any extra examples given by your classmates.

   _____

   _____

   _____

Drugs as medicines

Activity 2 *(continued)*

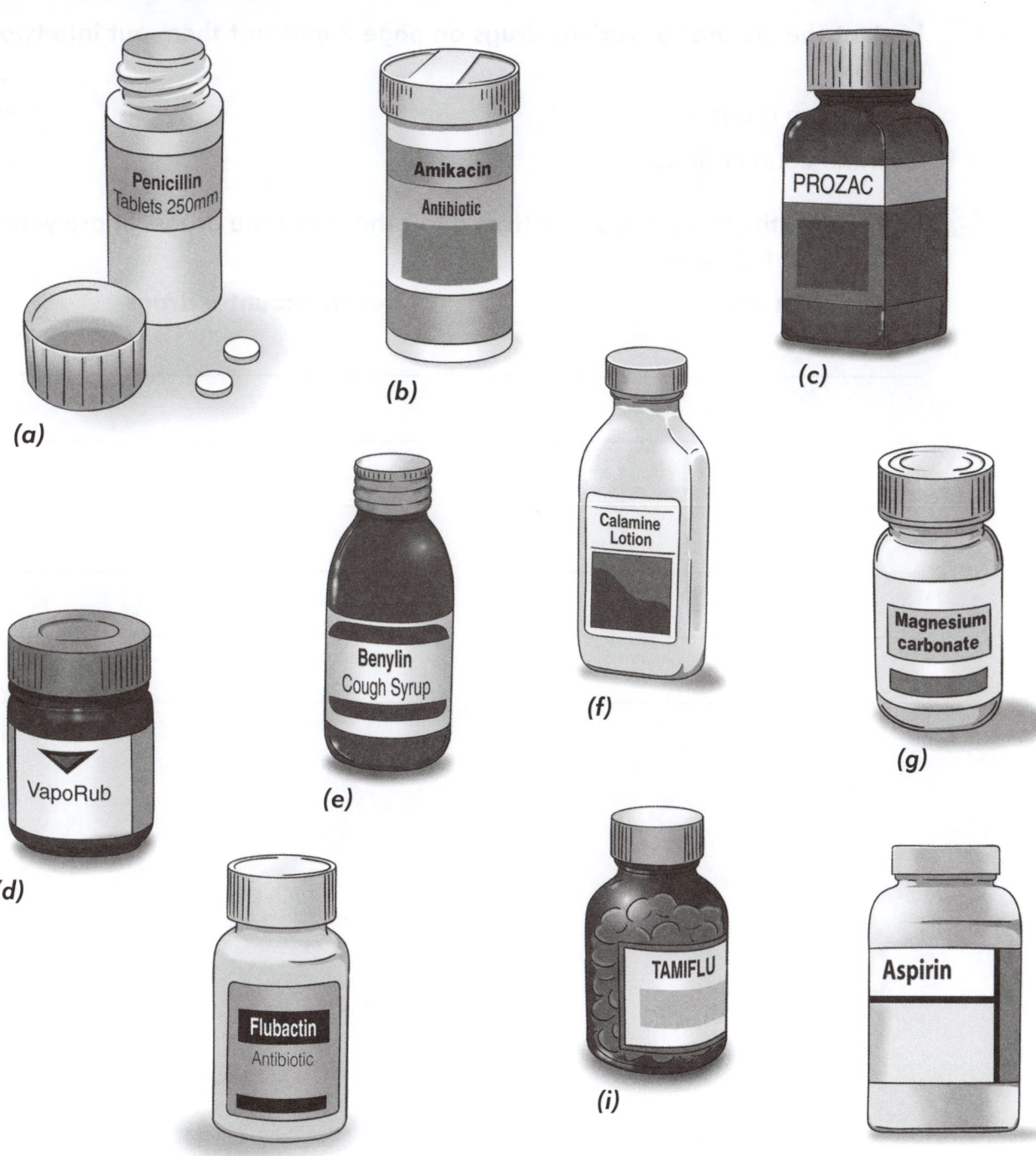

Chapter 1: Humans and animals

Activity 2 *(continued)*

4 Look at the pictures of various drugs on page 7 and sort them out into two groups:

   a Prescription drugs

   b Over-the-counter drugs.

5 Write down the two groups of letters *a* to *j* that label the drugs. Share your groups with the class.

Prescription drugs	**Over-the-counter drugs**

Drugs as medicines

Activity 3: Finding out more about drugs used as medicines

You will need: drug packets and bottles, a ruler and a pen or pencil.

1. Use the table below to record information about the drugs used as medicines.

   If you need to, use a ruler and pencil to make more rows.

2. Copy the information from the packets and bottles into the table.

3. Decide how you will present the information you have collected.

4. Sort the drugs out in some way. For example, create:

   a groups for each type of drug use

   b groups for each type of warning.

5. Display the drug containers.

Drug	Use	Dosage	Expiry date	Side effects	Warnings
Name:					
Name:					
Name:					
Name:					

Chapter 1: Humans and animals

Activity 3: Finding out more about drugs used as medicines *(continued)*

6 Share your findings with the class.

    a Discuss what you have all found out about the proper use of drugs as medicines.

    b Design a poster or web page about the safe use of drugs as medicines. Draw your ideas here:

Drugs as medicines

Activity 4 📖 12

You will need: a pen or pencil.

1 Discuss with your group:

    a 'What are the benefits of drugs?'

    b 'What are the harmful effects of drugs?'

2 Keep notes of your group's answers to these two questions.

Notes:

Benefits of drugs	Harmful effects of drugs

3 Share your group's answers with the class.

Chapter 1: Humans and animals

Activity B: One family's medicine cabinet

You will need: a pen or pencil.

Here is some data about the drugs found in the medicine cabinet of one family. Use the data to draw one bar chart with four columns to show these four types of drug:

A Over-the-counter liquid medicines
B Prescription tablets
C Over-the-counter tablets
D Prescription liquid medicines

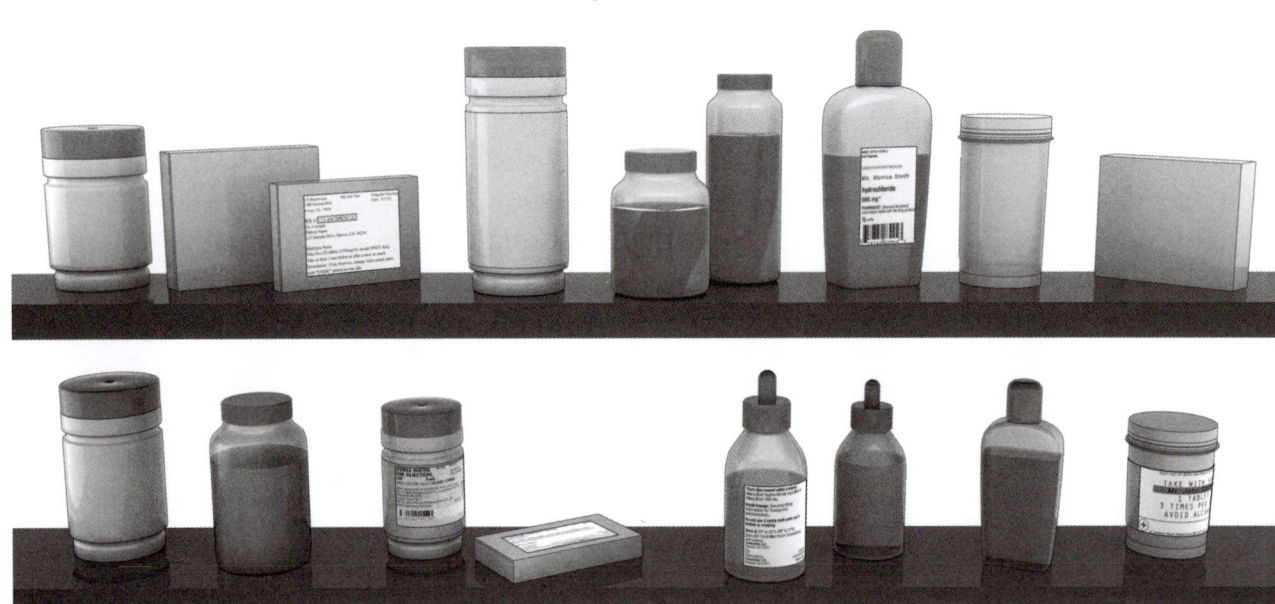

The cabinet had:

- seven types of liquid medicines – two of them were prescription drugs and the rest were over-the-counter drugs
- ten types of tablets – six of them were over-the-counter drugs and the rest were prescription drugs.

Drugs as medicines

Activity B: One family's medicine cabinet *(continued)*

Complete the bar chart below, using the data from the previous page.

Remember to label the vertical axis and name each of the bars with the correct letter: A, B, C or D.

Add the meaning of the letters in the key below the chart.

Don't forget to name your graph clearly.

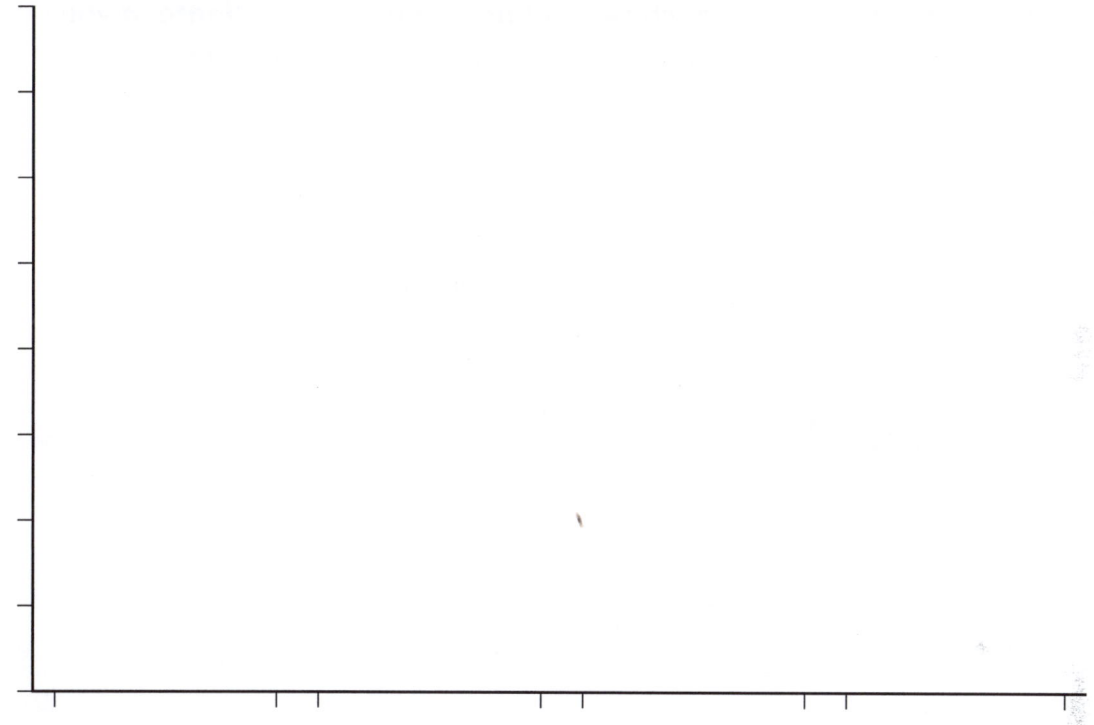

Key:

A _____   B _____

C _____   D _____

Chapter 2: Living things in their environment

Animals in their habitats

Activity 1: Investigate two different habitats 📖 14

You will need: two different habitats and a pen or pencil.

1 Discuss with your group which *two* habitats you will **investigate** in your locality. Try to choose two that are very different. Name them here:

Habitat 1: _____

Habitat 2: _____

2 Plan what you will look for and how you will record what you observe. Write down your plan here:

Plan

I will look for

I will record what I observe by

Animals in their habitats

Activity 1: Investigate two different habitats *(continued)* 14

 3 Go outside with your Workbook and pen, and investigate the chosen habitats.

a Collect enough information to be able to give a clear and full description of the animals and their habitats to the class.

 b Describe what each habitat is like.

Habitat 1: _____

- Is it hot and dry? _____

- Is it shady? _____

- Is it wet? _____

- Is it bare soil, rock or sand? _____

- Is it in or near water? _____

- Are there plants, and are they close together or spread out?
  _____

What else can you observe?
_____
_____

What animals can you see?
_____
_____

Draw the habitat here:

Chapter 2: Living things in their environment

Activity 1: Investigate two different habitats *(continued)* 📖 14

Habitat 2: _____

- Is it hot and dry? _____
- Is it shady? _____
- Is it wet? _____
- Is it bare soil, rock or sand? _____
- Is it in or near water? _____
- Are there plants, and are they close together or spread out?
  _____

What else can you observe?
_____
_____

What animals can you see?
_____
_____

Draw the habitat here:

Animals in their habitats

Activity 1: Investigate two different habitats (continued)

4 Return to class.

    a Discuss with your group the information you have collected.

    b Decide how you will present your descriptions to the class.

5 Present your descriptions and then answer questions from others in the class.

Each of the animals in the pictures above is shown in its natural habitat.

Chapter 2: Living things in their environment

17–18 Use these words to complete the sentences below:

> tadpoles tiger hide feed warm cold frogs
> slide smooth water flowers snakes eggs
> birds penguins sea shape fish nest bees
> pond deserts

1. The _____ lives in the jungles of India. Its stripes help it to _____ when it is hunting.

2. _____ live in large numbers in the ____. They swim and _____ together. Their _____ helps them to move easily through the _____.

3. The Antarctic is a very _____ habitat so the _____ and other animals have to have a way of keeping _____. The feathers protect the _____ from the cold.

4. Some _____ live in very hot, dry places such as _____. Their skin is dry and it is very _____ which helps them to _____ over the sand and rocks.

5. _____ and other insects visit _____ to collect food. They must use it for themselves or take it back to the _____ to feed the young.

6. A _____ is a good place for _____ because they must have fresh water to lay their _____ in. Adults can come out of the water, but the eggs and _____ must live in it as they grow.

Animals in their habitats

Activity C

You will need: a pen or pencil.

1 Look at the pictures above. *Think* about why the animals do not fit in those habitats.

   a What makes the habitats unsuitable for those animals?

   b What do the animals need in their habitats to stay alive?

2 Write a sentence about each animal, explaining why they cannot live in the habitats shown in the pictures.

   a The fish

   _____

   _____

   b The snail

   _____

   _____

   c The butterfly

   _____

   _____

   d The frog

   _____

   _____

Identification keys

Activity 2: Sorting and identifying animals by their features 19 – 20

You will need: a pen or pencil.

1. Discuss with your group what you remember from Stage 3 about grouping animals, using their simple features – for example, body covering or colour.

2. Look at the animals shown in the pictures above and discuss how they could be sorted.

Identification keys

Activity 2: Sorting and identifying animals by their features *(continued)* 📖 20

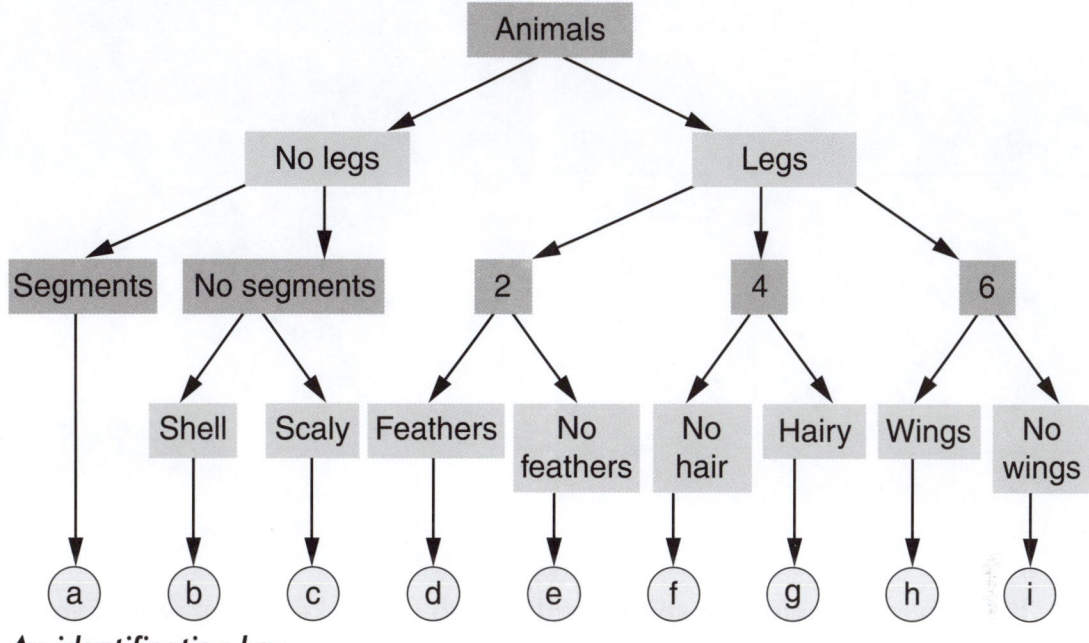

An identification key

3 You can use the **key** above to identify the animals in the pictures on page 20.

4 When you have identified the animals, write their names beside the correct letters.

a _____

b _____

c _____

d _____

e _____

f _____

g _____

h _____

i _____

5 Share your **results** with the class.

Chapter 2: Living things in their environment

Activity 2: Sorting and identifying animals by their features *(continued)* 📖 21

💬 **6** If you had to include a spider and a crab in the key, what would you add to it?

    a Discuss your ideas with the group. Write down your group's ideas here:

    _____

    _____

    _____

    _____

    b When you have an answer, share it with the class.

7 Choose another animal and ask your group to fit it into the key.

    Animal name: _____

    How does it fit into the key?

    _____

    _____

✏️   Share your group's ideas with the class.

Human activity and the environment

Activity 3 📖 22

You will need: an area outside and a pen or pencil.

1 Go outside with your group and choose a place where you can do a survey of litter on the ground. It might be inside or outside the school playground.

2 Keep a tally of each kind of rubbish you find (for example, plastic bags or drinks cans).

Write in the box below the names of each kind of rubbish you find, and write the tally next to each name.

Tally

3 a Use the data to make a bar graph.

Which kind of rubbish did you find the most of? Use that number when deciding how tall to make your vertical axis. Then write the names of the rubbish types under each column on your horizontal axis. Don't forget to give your graph a title.

Chapter 2: Living things in their environment

Activity 3 (continued) 22 – 23

3 b Display your group's graph with those from other groups.

4 Compare what the groups have found. Try to **explain** what the graphs show.

5 Make a poster with a message about how and why rubbish should be disposed of properly.

Draw your ideas here first:

Display your posters around the school.

Identification keys

Activity 4: What are the effects of litter?

You will need: a pen or pencil.

1 Look at the pictures and discuss what effects such things have on:
- a people
- b other living things
- c non-living parts of the environment.

Chapter 2: Living things in their environment

Activity 4: What are the effects of litter? *(continued)*

2. Write lists of all the effects that the group can think of.

3. Display your lists under the headings:

People

_____

_____

_____

Other living things

_____

_____

_____

Non-living things

_____

_____

_____

Identification keys

Activity 5 26

You will need: bins and a pen or pencil.

 1 Discuss with your group what materials you will **recycle** at school.

2 Make bins for each of the chosen materials. Put a label on each one. Write down a list of the labels you will use here:

 3 Write instructions to tell the class how to use the bins. Display the instructions near the bins.

How to use the bins

Chapter 2: Living things in their environment

Activity D

You will need: a pen or pencil.

1. Think of all the materials that can be recycled from the waste in your home. Write their names here:

   _____

   _____

2. Choose four of those materials. You will tally how many of each material you use over a week at home. Record each item in the table below.

Material for recycling	Tally of the items over a week							Totals
	Mon	Tues	Wed	Thurs	Fri	Sat	Sun	
1 _____								
2 _____								
3 _____								
4 _____								

3. When the table is complete, use the data to write four sentences about what you have found out.

   a _____

   b _____

   c _____

   d _____

Chapter 3: States of matter

Solids, liquids and gases

ice

clouds

rain

snowy peaks

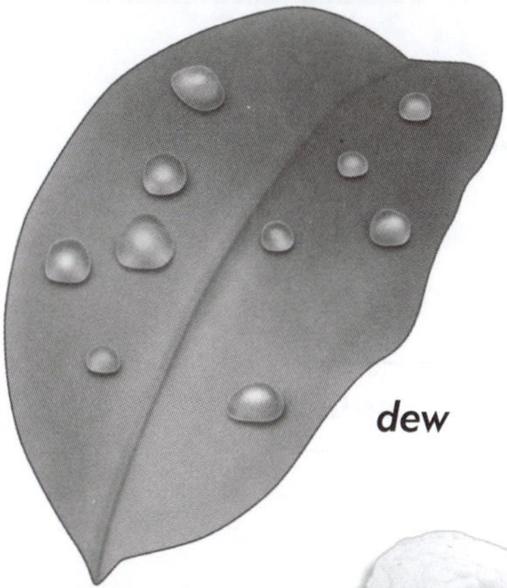

dew

river

hailstones

boiling water

Chapter 3: States of matter

Activity 1: Identifying water in its three states 📖 29

You will need: a pen or pencil.

 1 Look at the pictures on page 29.
   a Choose the ones that show water.
   b Tell the class which ones you have chosen.

2 Complete the table below:

Solid	Liquid	Gas

   a Identify the form of water in each picture.
   b Record your answers in the columns under the correct headings.

3 Complete these sentences.

Here are the words you will need:

> liquid   states   gas   water   solid

(You will need to use some words more than once.)

Water is found in three _____.

They are called _____, ____ and _____.

The _____ state is called ice.

_____ is the _____ state.

The ____ state is called water vapour.

30

Heating and cooling materials

Activity 2: Experiment to test the speed of drying pieces of cloth 32 – 33

You will need: water, pieces of cloth, scissors and a pen or pencil.

1 Investigate the drying of two pieces of cloth.
   a Plan how you will investigate the drying of two pieces of wet cloth.
   b Make it a **fair test** so that you will be able to compare the results.

2 How will you treat the two pieces of cloth?
   a Discuss with your group how you will treat the two pieces so that one dries faster than the other. Think about washing put out to dry.
   b **Predict** which piece will dry faster and write down your prediction.

Plan

I will investigate the drying of two pieces of wet cloth by

It will be a fair test because

I predict

Chapter 3: States of matter

Activity 2: Experiment to test the speed of drying pieces of cloth (continued)

3 Decide how you will **measure** the dryness of the cloths.

Keep a record of what you do, especially the time when you start and finish the investigation. Record this in a table.

Draw the table here using the column headings: **Time started**, and **Time finished**

Don't forget to add a column for how you measured the dryness of the cloths.

4 Discuss and compare the results.

  a Record what has happened to the cloths.

  _____

  _____

  b Discuss the results with your group.

  c Compare them and come to a conclusion based on your **evidence**.

  _____

  _____

  d Compare the results with your prediction.

  _____

  _____

5 Share your results with the class.

Heating and cooling materials

Activity E

You will need: a pen or pencil.

1 Four groups – A, B, C and D – did Activity 2 to find out which piece of cloth dried most quickly.

They were careful to make it a fair test.

2 To make it fair all the groups used:

_____

_____

_____

On the lines above write three things which they all kept the same.

3 They chose to weigh the cloths to find out which one had dried out most quickly. They weighed the cloths. When the test was over, they drew bar charts and compared their results.

4 Here are the four bar charts of their data.

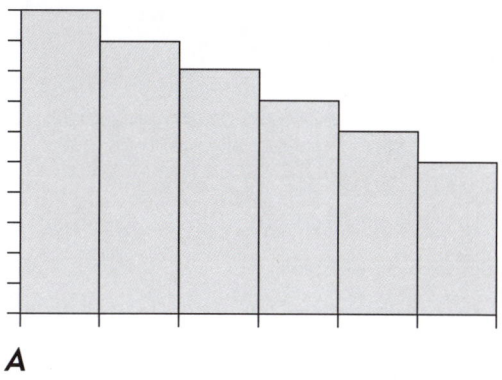

A

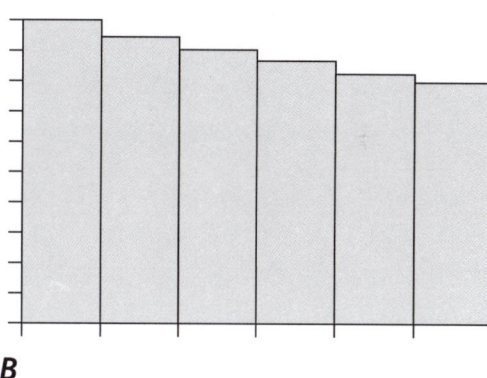

B

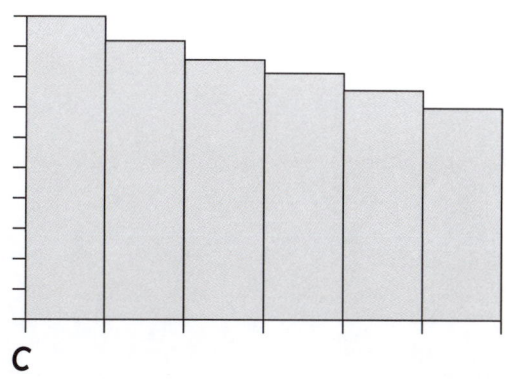

C

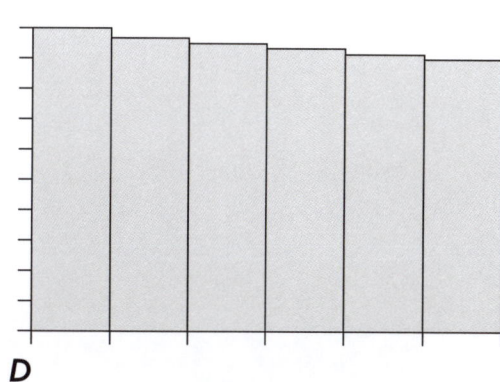

D

Chapter 3: States of matter

Activity E *(continued)*

5 Look at the charts and match the results to the four ways the groups treated their pieces of cloth. Write the chart letters in the spaces below:

- The cloth was hung out in the sun: Chart ☐
- The cloth was screwed up in a ball and left in the shade: Chart ☐
- The cloth was screwed up in a ball and left in the sun: Chart ☐
- The cloth was hung out in the shade: Chart ☐

6 Explain how you know which cloth was hung out in the sun.

_____

_____

_____

Heating and cooling materials

Activity 3: Investigating a material's changes of state 📖35

You will need: a heat source; either wax, chocolate, butter or margarine; and a pen or pencil.

1 Choose which material you will investigate: wax, chocolate, butter or margarine.

> ⚠️ **WARNING:** Take care when using the heat source to change the state of your material.

2 How will you change the state of your chosen solid material?

   a Discuss how you will do it with your group.
   b Write down what you plan to do.

Plan

I will change the state of my material by

3 Can you change the material back to its original state? How?

   a Discuss how you will **reverse** it with your group.
   b Write down what you plan to do.

Plan

I will reverse the state of my material by

Chapter 3: States of matter

Activity 3: Investigating a material's changes of state (continued)

4 Keep a record of what you do.

   a Draw a picture of your chosen material before you try to change its state.

   Before changing the _____

   b Record your observations of changes as they happen.

5 When the substance has changed from the solid state you started with, keep a record of what happens.

   a Make a second drawing of its new appearance and note any other differences. Use labels as well as writing a description of the change.

   After changing the _____

Heating and cooling materials

Activity 3: Investigating a material's changes of state *(continued)*

5 b Now try to reverse the change so that you return the material to its solid state.

c Record observations of changes as they happen. Use labels as well as writing a description of the change.

Change reversed

6 Discuss with your group what you conclude about why materials change their states. Write down the group's conclusion.

_____

_____

7 Share your drawings and conclusion with the class.

Steam

Activity 4: Investigating what happens to water when it boils 📖

> **You will need:** a pen or pencil.

Your teacher will use a kettle, a plate and a bowl to demonstrate another process.

> ⚠️ **WARNING:** When the water boils, take care not to get too close, or to put your hand near.

1 Look carefully at the spout and see if you can observe the 'gap' between the cloud and the spout.

2 Now look at the cloud.

   a What do you think it is – gas or liquid?

   _____

   b Tell the class what you think.

3 Watch carefully as the teacher holds the cold plate with the cloth and puts it into the invisible steam.

   The bowl is standing below the point where the plate is being held.

Steam

Activity 4: Investigating what happens to water when it boils *(continued)* 40

4 a What do you observe on the plate?

_____

_____

b What do you observe in the bowl?

_____

_____

c Try to explain your observations to the class.

---

41 Complete these sentences.

Here are the words you need:

> air   see   water   clouds   vapour   ran
> condenses   drops   dripped   cools   liquid

I saw _____ of _____ on the plate. They _____ down the plate and _____ off into the bowl.

Water _____ in the air _____ down as the _____ goes higher up.

As it cools, it _____ into tiny _____ of _____ water.

We can _____ the drops of _____ as _____ in the sky.

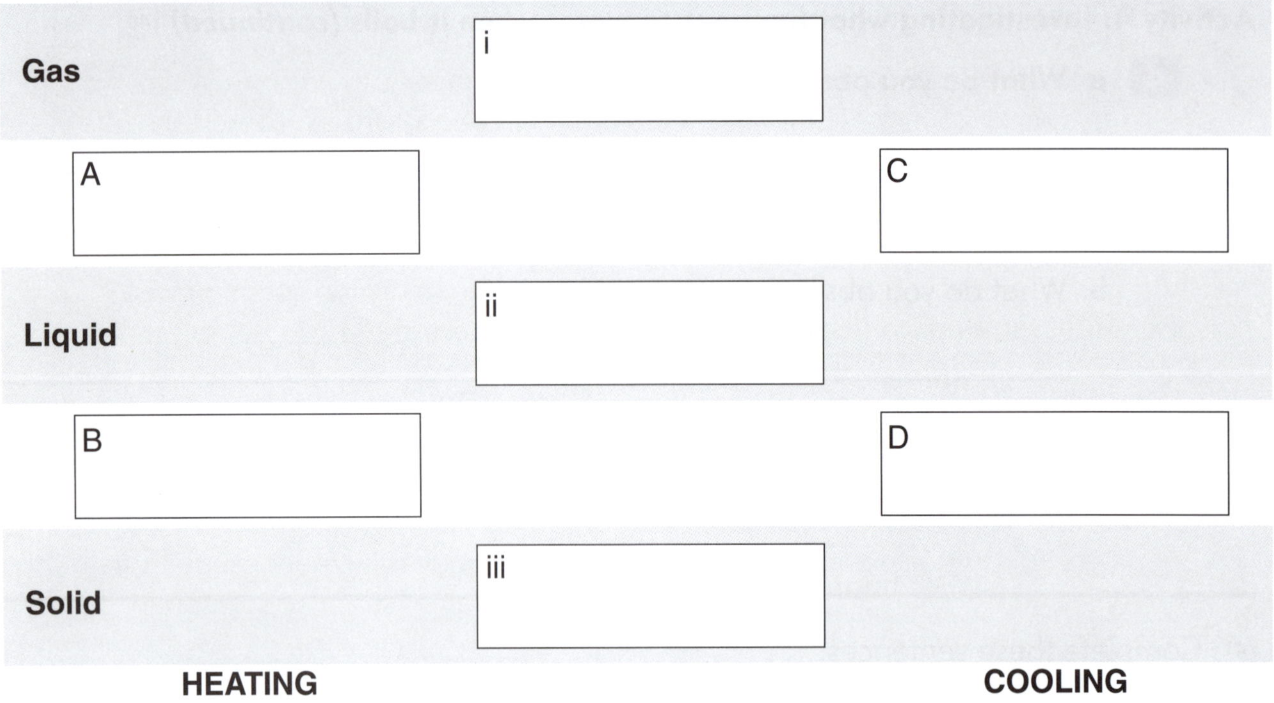

Chapter 3: States of matter

Gas	i	
	A	C
Liquid	ii	
	B	D
Solid	iii	

HEATING　　　　　　　　　　　　　　　　COOLING

Activity F

You will need: a pen or pencil.

1 Look at the diagram above showing the changes of state for **water**.

　　The column on the left shows the changes when **heating**.

　　The column on the right shows the changes when **cooling**.

Activity F *(continued)*

2 The middle column on page 40 has boxes for you to write in the names of the **three states of water**. Fill in boxes **i**, **ii** and **iii**.

3 Think of what happens when **solid** water is **heated**.
Write the name of this **process** in Box B.

4 Think of what happens when **liquid** water is **heated**.
Write the name of this **process** in Box A.

5 Think of what happens when **liquid** water is **cooled**.
Write the name of this **process** in Box D.

6 Think of what happens to water in the state of a **gas** when it is **cooled**.
Write the name of this **process** in Box C.

7 There are spaces for you to write in the names of the **three states** in the column on the right.

8 Add arrows to the three columns to show the direction of the changes.

Chapter 4: Sound

Sound making and measuring

Activity 1: Where do sounds come from? 42

You will need: an area outside and a pen or pencil.

1. Discuss with your group where sounds come from.

2. Go outside.
   a Listen for sounds.
   b Name as many sources of sounds as you can.

3. Write down each of the sounds you hear in the writing box below.

4. Share them with the class.

Sound making and measuring

Activity 2 📖 44

(a)
(b)
(c)
(d)
(e)
(f)
(g)
(h)
(i)
(j)
(k)
(l)
(m)

43

Chapter 4: Sound

Activity 2 (continued) 44

You will need: a pen or pencil.

1 Look at the pictures on page 43.
   a Discuss them with your classmates.
   b Are they all sources of sound?
   Yes   No

2 Sort them into groups:
   a natural sources of sound
   b artificial sources of sound
   c non-sound sources.

3 Use the letters on the pictures to record your groups.
   a Natural sources of sound:
   _____

   b Artificial sources of sound:
   _____

   c Non-sound sources:
   _____

4 Share your answers with the class.

Sound making and measuring

Activity 3: How are sounds produced? 45

You will need: a wooden or plastic ruler and a pen or pencil.

1. Take a ruler or flat strip of wood or plastic.
   a. Place it on the edge of the desk so that more than half of it sticks out over the edge.
   b. Hold it down firmly.

2. Flick the free end of the ruler. Repeat this several times and listen carefully.

3. Observe what happens to the ruler when you flick it – use your senses of sight, hearing and touch.

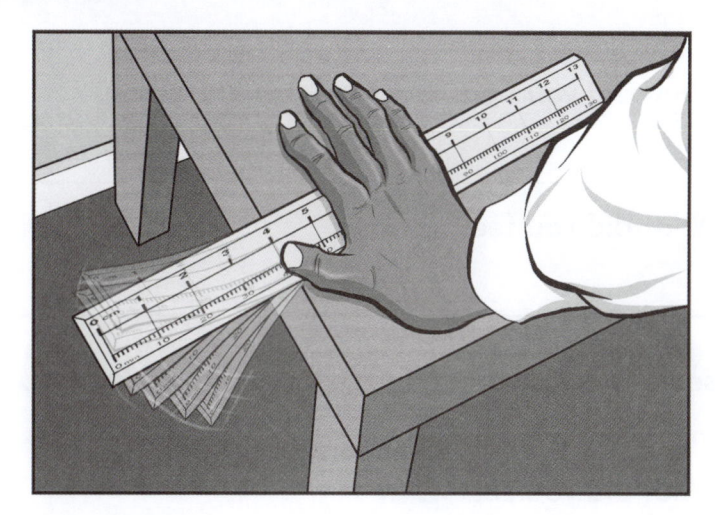

4. Record your observations.
   - What did you see 👁 ?

   _____

   - What did you hear 👂 ?

   _____

   - What did you feel ✋ ?

   _____

5. Discuss your results with the group.
   a. Come to a conclusion about how the sounds were produced.

   _____

   b. Share your conclusion with the class.

Chapter 4: Sound

Activity 4 📖 47 – 48

> **You will need:** an object that you can blow into or across to make a sound and a pen or pencil.

1 Choose an object that you can blow into – a **wind instrument**, tube, pipe – or blow across the open top of a glass or plastic bottle.

2 Try to make sounds with your chosen object.
  a Ask your group to observe the sound-making – look, listen and touch.
  b Write down a record of what the group observes.

  • What did you see 👁 ?
  _____

  • What did you hear 👂 ?
  _____

  • What did you feel ✋ ?
  _____

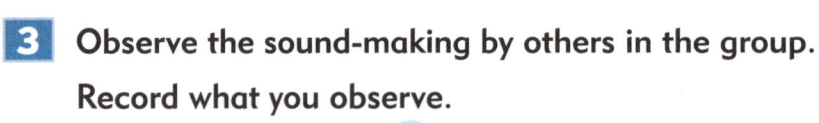

3 Observe the sound-making by others in the group.
  Record what you observe.
  • What did you see 👁 ?
  _____

  • What did you hear 👂 ?
  _____

  • What did you feel ✋ ?
  _____

4 Compare the records made by the group members.
  a Try to reach a conclusion based on the evidence of your observations.
  _____
  _____

  b Be ready to tell the class what you heard, what you saw and what you felt.

Sound making and measuring

Activity 5: Measuring sounds using a sound-level meter 48 – 49

You will need: a sound-level meter and a pen or pencil.

1 Handle the **sound-level meter** carefully.
 a Explore how it is switched on and off and how it is used to measure the **volume** of a sound.
 b Look at how you can see the measurement readings. They are in **decibels** (written **dB**).

2 When you are familiar with the meter, move around the room and outside.
 a Take measurements of sounds.
 b Record your observations in the correct units.

 Sound measurements:

3 Plan how your group will test making a sound louder and louder, using the meter to measure the level each time. It must be a fair test. When you have a plan, show it to your teacher.

Plan

I will make the sound louder and louder by

It will be a fair test because

47

Chapter 4: Sound

Activity 5: Measuring sounds using a sound-level meter *(continued)* 49

4 Do your test and record the readings from the sound-level meter each time you measure the sounds you make.

Sound name: _____

Test 1 description: _____ Decibels: [ ]

Test 2 description: _____ Decibels: [ ]

Test 3 description: _____ Decibels: [ ]

5 When you have finished the test, look at your data.

  a Discuss them with your group and come to a conclusion about the sounds you made.

  _____

  _____

  b If you can use the data to produce a bar chart or a table, then do it, ready to share with the class. Record it on page 49.

A table could be used, like this example.

Sound source e.g. dropping a book on the desk	Sound level (dB)
From 5 cm	x
From 10 cm	y

Or a bar chart could be used like this example.

(Bar chart: Sound level (dB) vs Height above desk (cm), bars at 1, 2, 3, 4, 5 increasing in height)

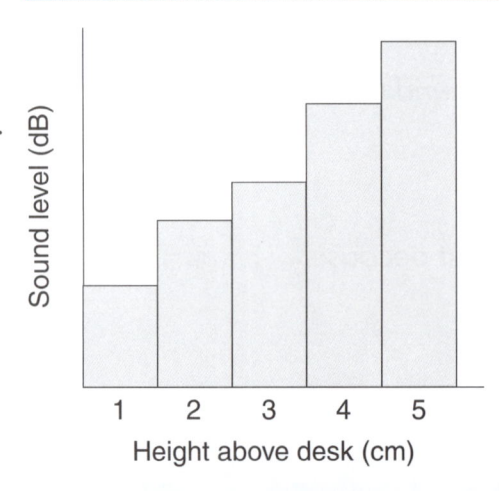

Sound making and measuring

Activity 5: Measuring sounds using a sound-level meter *(continued)*

If you can use the data to produce a bar chart or a table, then do it, ready to share with the class. Record it here:

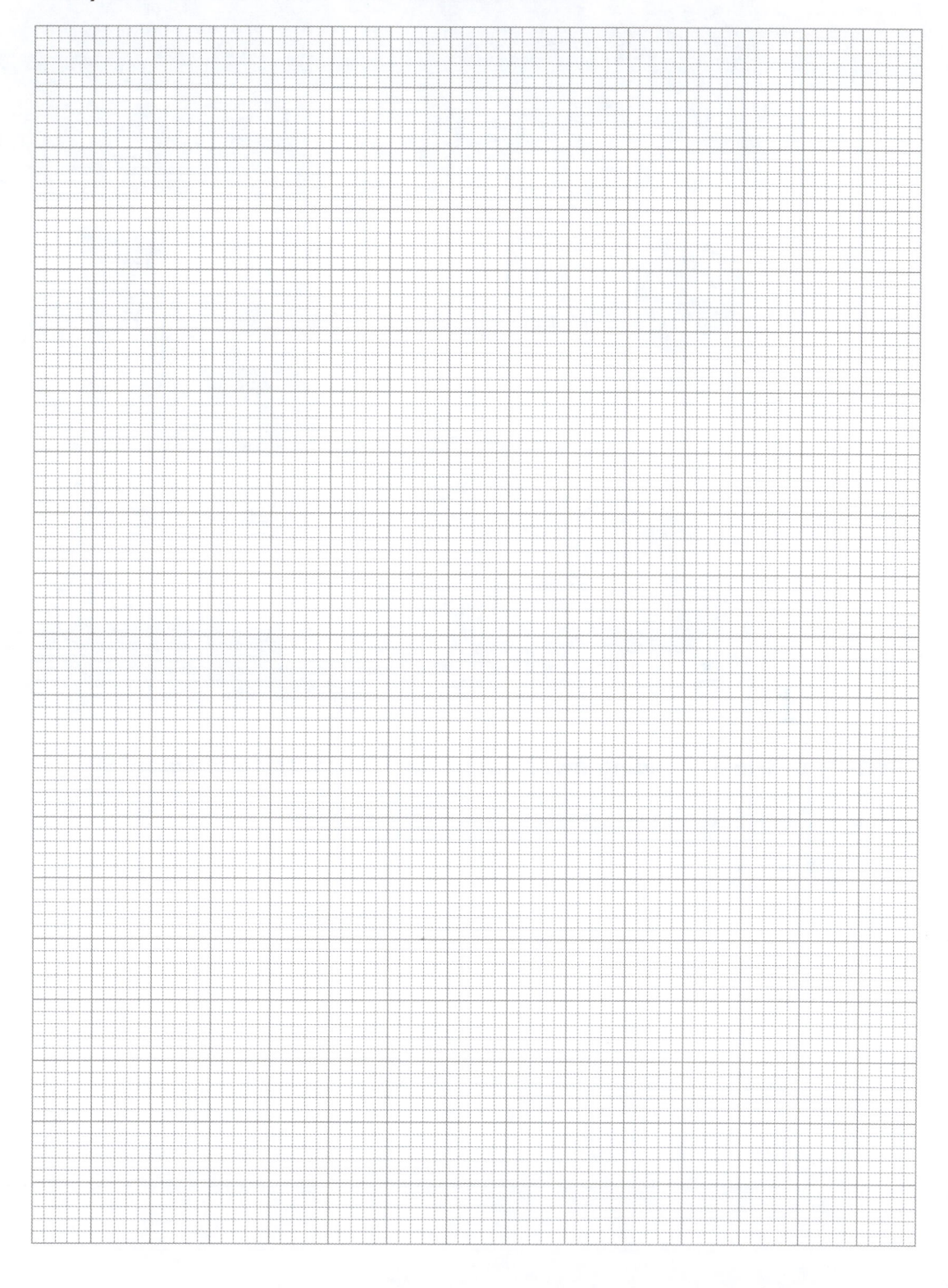

Sound travelling

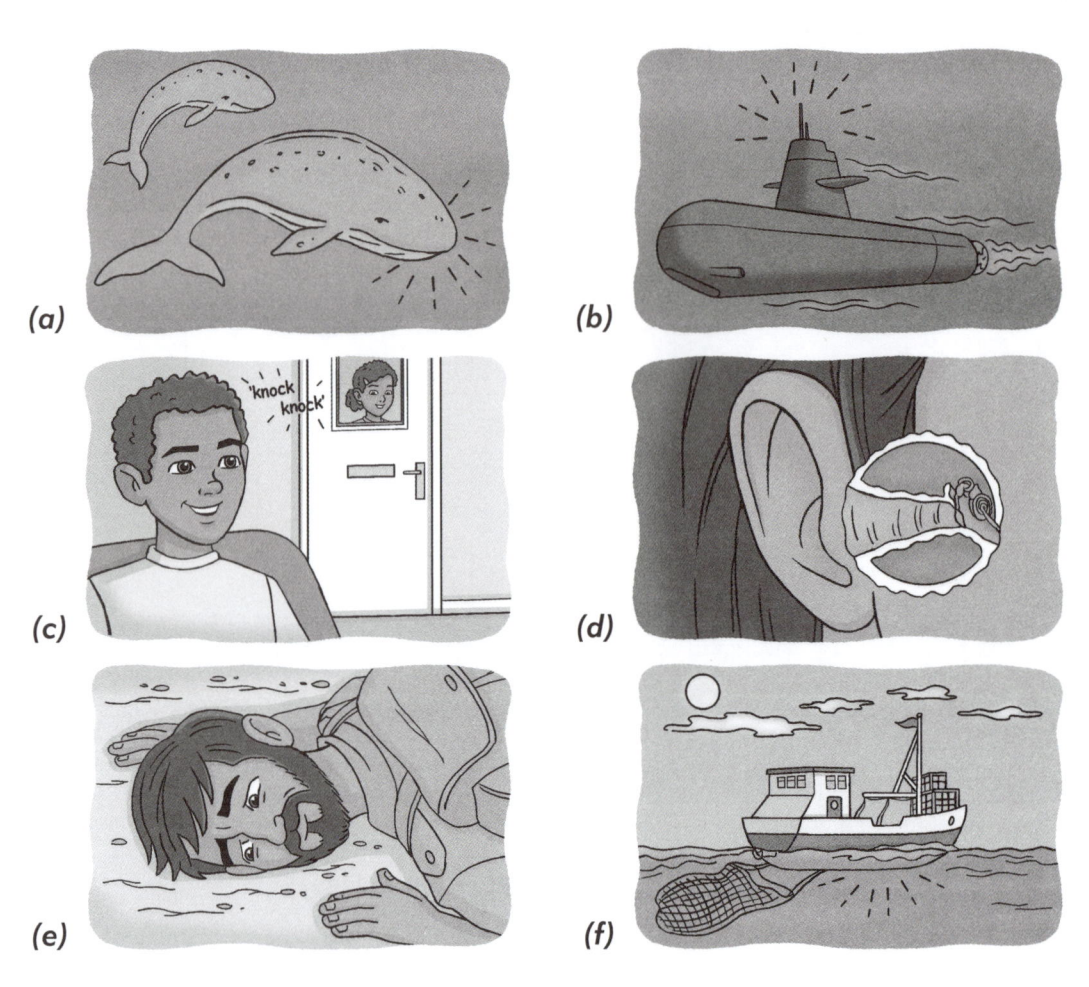

(a)
(b)
(c)
(d)
(e)
(f)

Activity 6: Understanding more about the transmission of sound 📖 52

You will need: a pen or pencil.

🔍 **1** Look at the pictures.

💬 a Discuss with your group what they show.

_____
_____
_____
_____
_____
_____

Sound travelling

Activity 6: Understanding more about the transmission of sound *(continued)*

1 b Try to work out what they have to do with the transmission of sound.

_____

_____

2 For each situation, decide if sound is being transmitted and, if it is, what is it travelling through – a solid, a liquid or a gas? Circle your answers below.

a Is sound being transmitted?   Yes   No

It is travelling through a solid / a liquid / a gas.

b Is sound being transmitted?   Yes   No

It is travelling through a solid / a liquid / a gas.

c Is sound being transmitted?   Yes   No

It is travelling through a solid / a liquid / a gas.

d Is sound being transmitted?   Yes   No

It is travelling through a solid / a liquid / a gas.

e Is sound being transmitted?   Yes   No

It is travelling through a solid / a liquid / a gas.

f Is sound being transmitted?   Yes   No

It is travelling through a solid / a liquid / a gas.

3 Share the group's ideas with the class.

Chapter 4: Sound

Activity 7: What happens to sounds when they travel through a solid? 53 – 54

You will need: a desk or table, a stick and a pen or pencil.

1 Work with a partner for this activity.
  a Sit at opposite ends of the desk or table.
  b Cover one ear with a hand and put your other ear close to the desk top, but *not* touching it.
  c Ask your partner to lightly scratch the desk top with a pencil or stick.
  d Now put your ear closer to the desk so that it *is* touching the desk top.
  e Ask your partner to repeat the scratching.

2 Swap with your partner and repeat the activity.

3 Discuss what you both heard and come to a conclusion.

Sound travelling

Activity 8: What happens to sounds when they travel through a liquid?

You will need: water; a balloon; a ticking watch, clock or timer; and a pen or pencil.

1 **Work with a partner on this activity.**

   a Ask your partner to hold the watch, clock or timer alongside your head so that you are just able to hear the sound of the ticking.

   b Listen carefully to what it sounds like. What can you hear?

   _____

2 **Fill the balloon with water and place it beside your head so that it touches your ear.**

   a Ask your partner to put the watch, clock or timer on the other side of the balloon, so that it is touching it.

   b Listen carefully and compare what you hear this time with the first time. What can you hear?

   _____

3 **Swap with your partner and repeat the activity.**

4 **Discuss the observations you both made and come to a conclusion about sound travelling through water.**

   _____

   _____

Materials preventing sound transmission

Some materials reduce the loudness of sounds or even completely stop the sound waves from travelling through them.

Activity 9: Which material will be best at preventing sound travelling through it?

📖 57 – 58

> **You will need:** a selection of materials to test, a sound source, a sound-level meter and a pen or pencil.

1 Discuss with your group which materials you think can reduce or prevent sound travelling from its source (its transmission).

2 Plan an investigation of how sound transmission can be prevented or reduced.

   a Make it a fair test, using at least *three* materials and a sound source.

   b Predict what you think the results will show.

   c Write down your prediction.

   Name the materials here:

   1 _____

   2 _____

   3 _____

Plan

I will investigate how sound transmission can be prevented or reduced by

It will be a fair test because

I predict

---

3 Decide how you will observe the reduction or prevention of the sound. What will you measure, and how?

   I will measure _____

   I will measure this by _____

Materials preventing sound transmission

Activity 9: Which material will be best at preventing sound travelling through it? (continued) 58

4 Decide how you will record your observations and prepare for the recording.

5 Compare the effects of the three different materials on the transmission of the sound.

   a Carry out the test in a fair way so that you will be able to compare the effects.

   b Record your measurements each time in the box below.

6 Compare your results.

   a Use them to make a conclusion about the three materials.

   _____

   _____

   _____

   b Compare your results with your prediction. What was different and what was the same?

   _____

   _____

7 Share your group's results with the class and compare them with those from other groups.

Pitch and loudness

Activity G

You will need: a pen or pencil.

1 A group used a sound-level meter to measure the volume of sound made by a person reading aloud from a book. The voice had been recorded and the volume control on the player was set at one position and not changed through the investigation.

2 The chart below shows the results collected by the group.

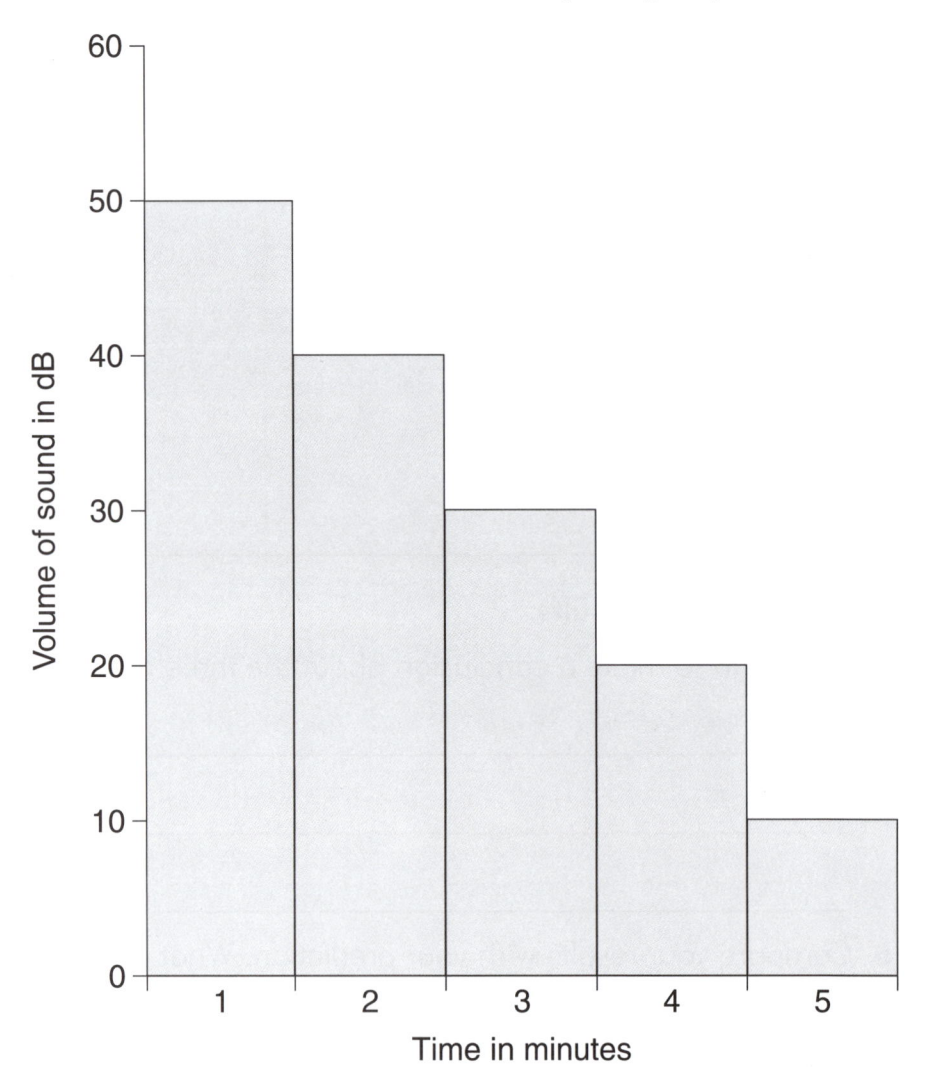

Pitch and loudness

Activity G (continued)

3 Look at the way the measurements gradually fell, even though the sound from the player was not being changed.

4 Think about *why* the sound-level meter readings went lower and lower. Write your explanation in the space below:

_____

_____

_____

_____

5 Now complete the bar chart below. It should show what happens to the volume of the sound when you do the *opposite* of what the groups did.

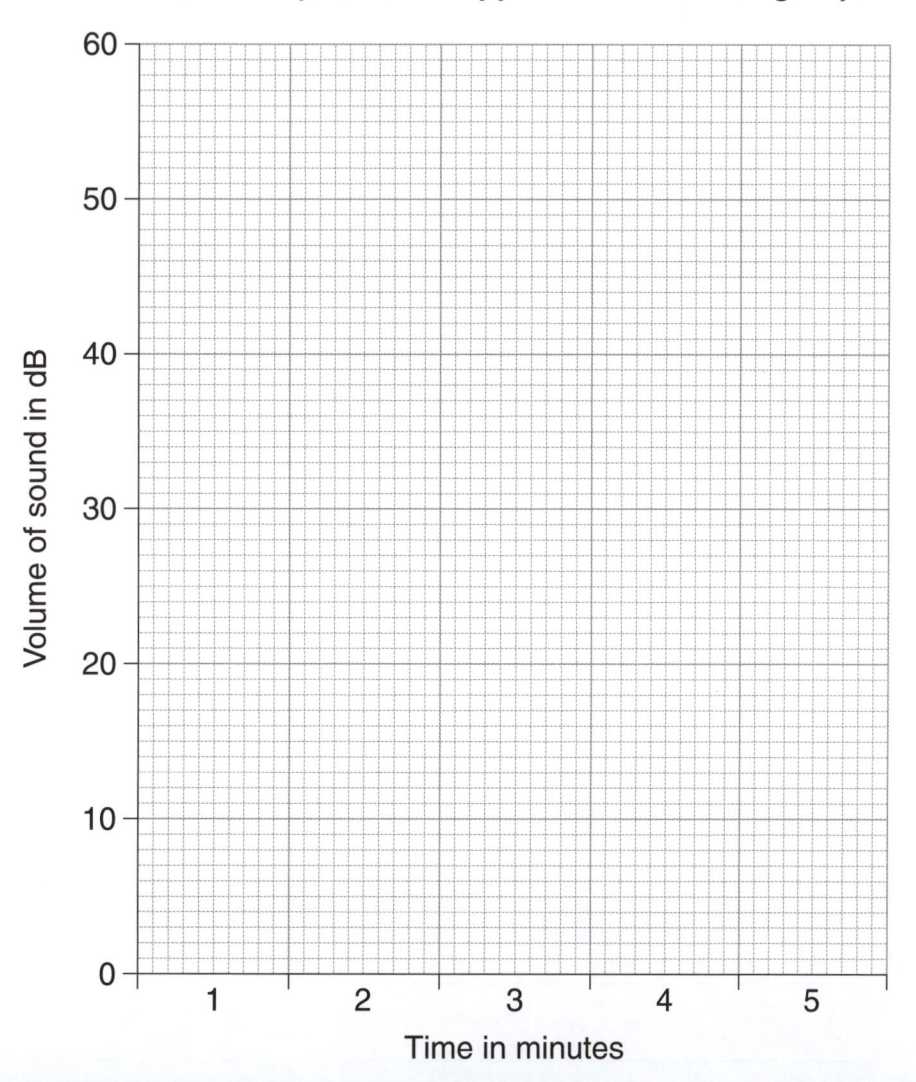

Volume of sound in dB (y-axis, 0–60)
Time in minutes (x-axis, 1–5)

Chapter 4: Sound

Activity H

You will need: six sound sources and a pen or pencil.

1 Choose six sources of sound at home, each one with a different pitch. Draw pictures of them in the boxes below.

2 Put them in order of pitch, from **1**: the lowest pitch, to **6**: the highest pitch.

3 Write the names of the sources under your drawings.

Pitch and loudness

Activity 10 |61|–|62|

You will need: a wooden or plastic ruler, a desk or table and a pen or pencil.

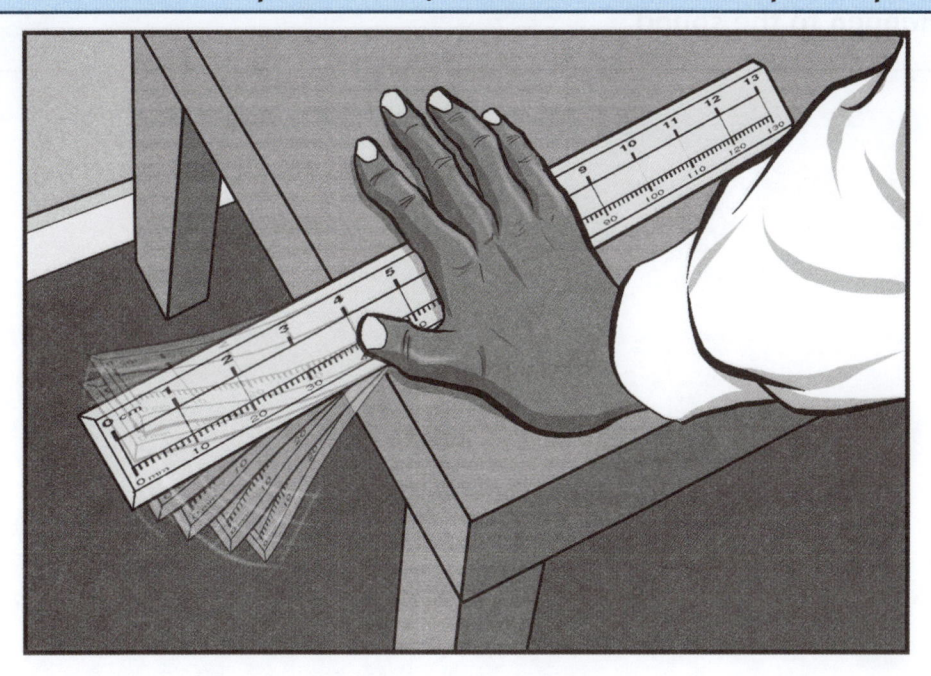

1 Take a ruler or flat strip of wood or plastic.
  a Place it on the edge of the desk so that more than half of it sticks out over the edge.
  b Hold it down firmly.

2 Flick the free end of the ruler. Repeat this several times and listen carefully.

3 Discuss with your group how you can change the pitch of the sound:
  a to make it a lower sound

  _____

  b to make it a higher sound.

  _____

Chapter 4: Sound

Activity 10 (continued) 📖62

4 Write down in a table what you will do and your prediction of what will happen to the sound.

5 Test the group's ideas and record in the table what happens to the pitch of the sound each time you try to change it.

6 When you have tested all the ideas, compare the results with the predictions and try to explain what you have observed.

   a Can you see a **pattern** in the results?

   _____

   _____

   b What can you say about the pitch of the sound and the length of the ruler sticking out from the desk?

   _____

   _____

7 Share your results and conclusions with the class.

Pitch and loudness

Activity 11 〔63〕

You will need: a rubber band and a pen or pencil.

1 Hold a rubber band over your thumb and forefinger and stretch it out.

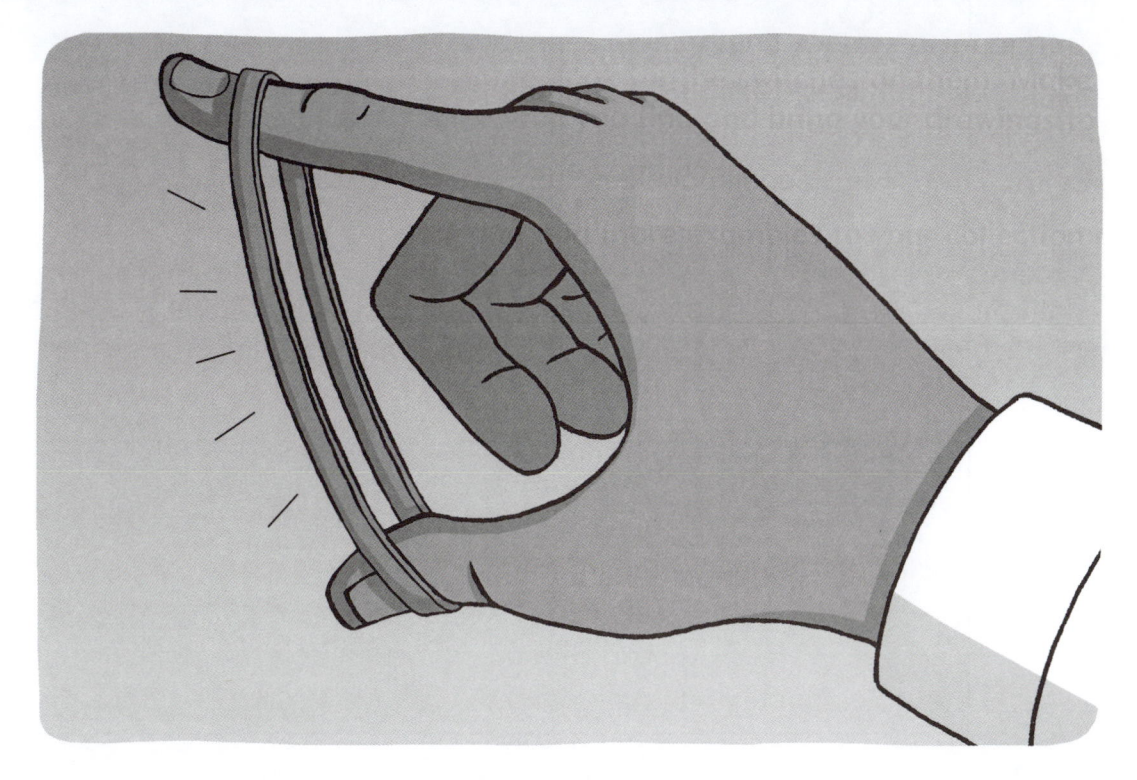

2 Pluck the rubber band with the other forefinger and listen to the sound you have made.

3 Discuss with your group how you can change the pitch of the sound:

  a to make it a lower sound

  _____

  b to make it a higher sound.

  _____

Chapter 4: Sound

Activity 11 *(continued)*

4 Write down in a table what you will do, and your prediction of what will happen to the sound.

5 Test the group's ideas and record in the table what happens to the pitch of the sound each time you try to change it.

6 When you have tested all the ideas, compare your results with your predictions and try to explain what you have observed.

   a Can you see a pattern in your results?

_____

_____

   b What can you say about the pitch of the sound and its link to how much you stretched the rubber band?

_____

_____

7 Share your results and conclusions with the class.

Pitch and loudness

65 Complete these sentences, using the words listed below:

> higher   longer   lower   shorter

a The pitch of the note gets _____ as the ruler gets _____.

b Making the ruler _____ makes the pitch _____.

c The pitch of the note gets _____ as the rubber band is stretched.

Now add three more sentences of your own about what you discovered when you tried to change the pitch of sounds:

d _____

e _____

f _____

Loudness (volume) is not the same as pitch. A sound can be made louder or softer without changing the pitch.

Sing a note loudly, then softly.

Clap your hands loudly, then softly.

Tap the desk loudly, then softly.

Did the pitch of the sounds change?

Activity 12 66

You will need: a rubber band, an empty match box or small plastic box, a sound-level meter and a pen or pencil.

1 Stretch a rubber band over a matchbox or other small container.

2 Pluck the band and listen to the sound it makes.

3 Discuss with your group how you can change the loudness of the sound to make it softer or louder.

Chapter 4: Sound

Activity 12 (continued)

4 Write down in the table below the group's ideas and predictions of what will happen.

Method of making sound	Predicted result	Sound level measurement

5 Test each of the ideas and listen carefully to the sounds you make.

6 Record the results in the table.

7 Compare the results with your predictions and try to explain what you observed.

_____

_____

a Look for patterns in your results and come to your conclusions.

_____

_____

b Share your results and conclusions with the class.

Musical instruments

Activity 13 [69]

You will need: a musical instrument and a pen or pencil.

1. Choose an instrument and play with it, exploring how you can change the pitch of its sounds.

2. Record the name of the instrument and the methods that you used successfully to change the pitch of its sounds.

    Name of instrument: _____

3. Choose a different type of instrument and carry out the same exploration.

4. Record the instrument's name and methods as before.

    Name of instrument: _____

5. Take a third type of instrument and repeat the activity.

    Name of instrument: _____

6. Compare your results.

    a. Come to a conclusion about how the pitch of notes from musical instruments can be changed.

    b. Can you see any patterns in your observations?

7. Share your results and conclusions with the class.

Chapter 5: Electricity and magnetism

Constructing circuits

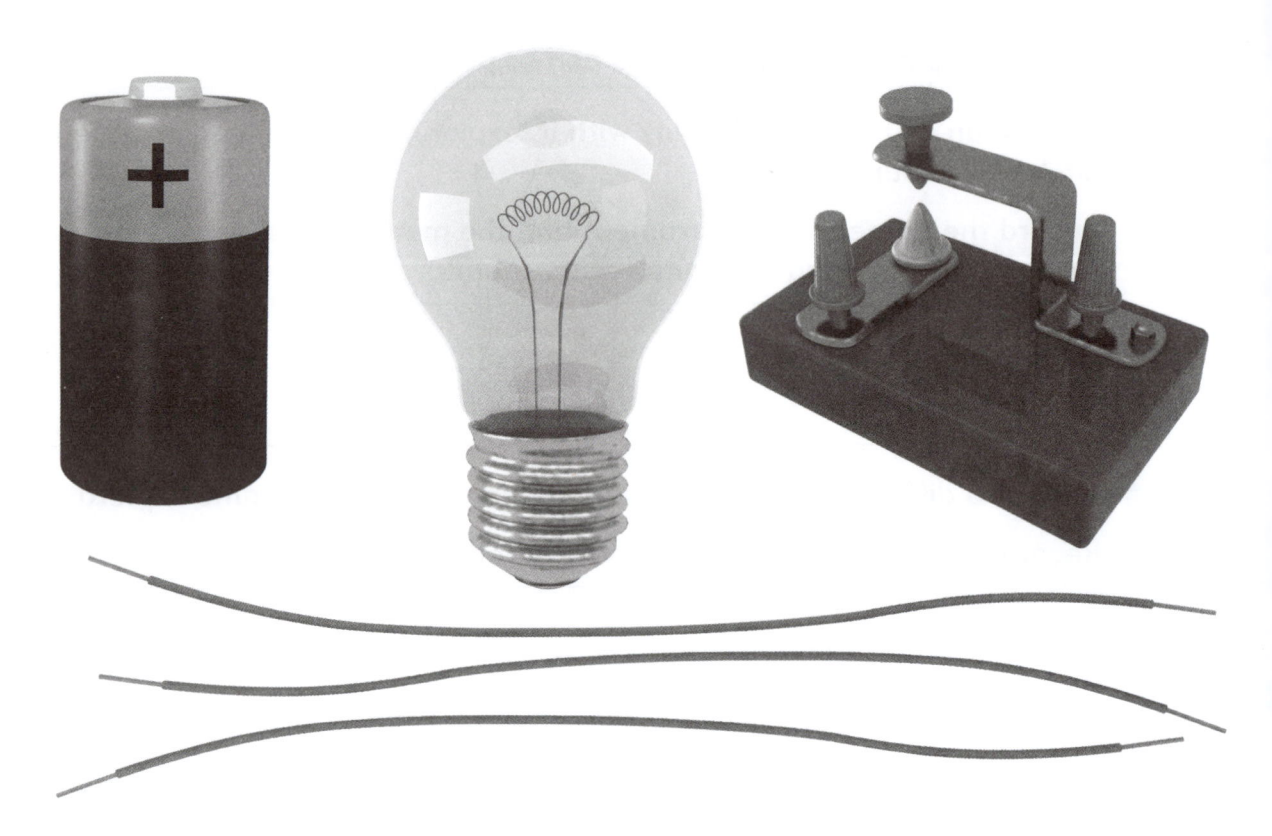

Activity 1 📖72

You will need: a battery (cell), a lamp, three insulated wires with bare ends showing, a switch and a pen or pencil.

1 Collect the items shown in the picture.

2 Try to arrange the six things in a complete **circuit** so that the lamp lights up and can be turned off.

3 Change the way you arrange them.

Constructing circuits

Activity 1 *(continued)* 📖 73

4 Record in simple drawings all those ways which make the lamp light up and allow you to switch it off.

Use labels on your drawings here:

5 Be careful to show which places on the **battery** and lamp are touching the wires.

Chapter 5: Electricity and magnetism

Activity 1 *(continued)* 73

6 Try the arrangement shown in the picture:

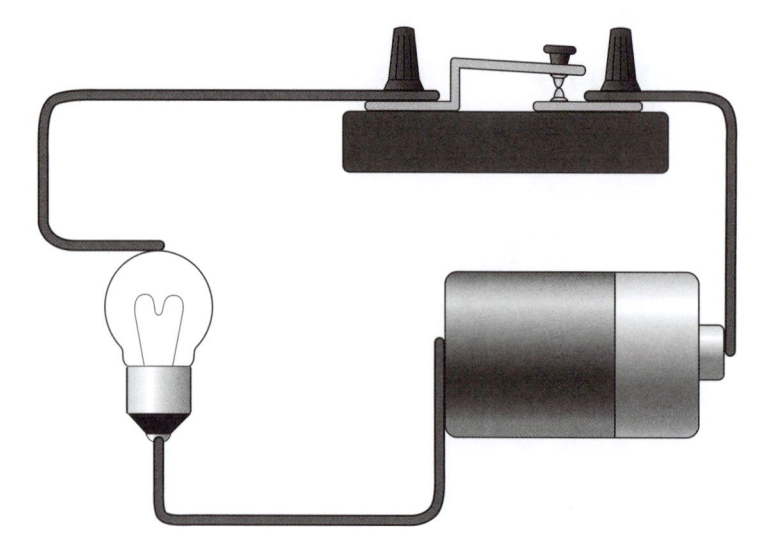

Does it make the lamp light up?

Yes   No

7 Share your results with the class.

Broken circuits

Activity 2 |75|

You will need: electrical items with switches and a pen or pencil.

1 If you have electrical items at home, look for the switches on them. Make a simple drawing of each kind of switch you find and bring your drawings to school.

2 Continue this activity in school and add more examples to your collection of drawings.

Drawings of switches in school

3 Display your drawings for the class to see and look at the drawings of other students.

Chapter 5: Electricity and magnetism

Activity 3 📖 76

You will need: a battery (cell), a lamp, three insulated wires with bare ends showing, a switch and a pen or pencil.

1 Rebuild a circuit that lights the lamp.

2 Explore how many ways you can break the circuit and make the lamp go out. Make a simple drawing of each broken circuit.

Circuit 1

Circuit 2

Circuit 3

3 Share your results with the class and explain why each circuit is broken.

Broken circuits

(a)

(b)

(c)

(d)

Which of these circuits are complete?

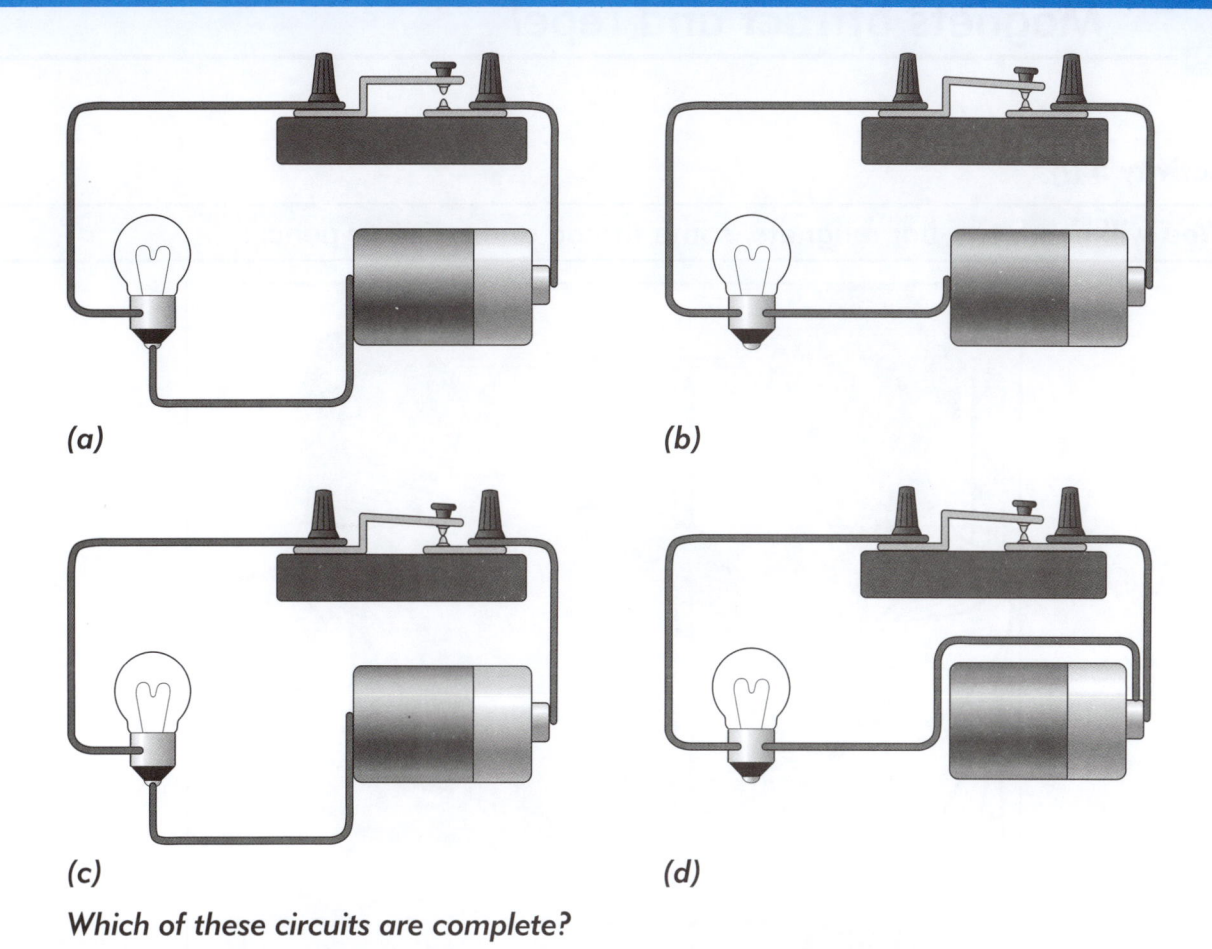

Activity 4

You will need: a pen or pencil.

1 Look at the drawings of the four circuits. Which circuits will light the lamp?

2 Discuss your ideas with your group. Write them here.

_____

_____

_____

_____

Be ready to explain your answers to the class.

Magnets attract and repel

Activity 5 📖 79

You will need: two bar magnets, some thread and a pen or pencil.

1 **Tie a thread to one of the bar magnets.**

   a Hold the bar magnet on the thread letting it swing freely.

   b When it is settled, bring another bar magnet slowly towards it – aiming the north pole of one towards the north pole of the other.

   c Write a note of what you observe in the box below. Add drawings if you wish.

Magnets attract and repel

Activity 5 *(continued)* 79 – 80

2 Repeat the action using the south poles of both magnets.

Write a note of what happens this time in the box below. Add drawings if you wish.

3 Now bring the south pole of one magnet slowly towards the north pole of the other.

Write a note of what you observe in the box below. Add drawings if you wish.

4 Lastly, bring the north pole of one magnet towards the south pole of the other.

Record what happens this time in the box below. Add drawings if you wish.

5 Share your results with the class.

Metals and magnets

Activity 6: Identifying magnetic materials 📖 81

You will need: a bar magnet, 10 different objects and a pen or pencil.

1 Draw a table here to record the results of the activity.
Use these column headings:

Name of object	Material	Magnet attracted	Magnet not attracted

2 Take a bar magnet and use it to test 10 different objects in the classroom and outside.

a For each object, can you feel the pulling **force** of the magnet (attraction), or not?

b Record the result for each one in the table, with a tick or a cross in the correct column.

Metals and magnets

Activity 6: Identifying magnetic materials (continued) 📖 82 – 83

3 Look at the results.

   a Come to a conclusion about the *materials* attracted by the magnet.

   b Write down your conclusion.

_____

_____

_____

_____

   c Share it with the class.

4 Test your idea about **magnetic** materials on eight more objects in the room.

   a Write a sentence about the results.

_____

_____

Chapter 5: Electricity and magnetism

Activity 6: Identifying magnetic materials *(continued)*

4 **b** Compare your answer to 4a with your conclusion from the first test you did.

  c Do your conclusions agree, or have you changed your mind about which materials are magnetic?

  _____

  _____

  d Tell the class what you now think.

5 Sort out these materials into two groups: a **magnetic** group and a **non-magnetic** group.

> glass   plastic   iron   paper   gold   soil   copper
> cloth   rubber   steel   brass   stone   pottery   bone
> silver   tin   wood

The magnetic group   **The non- magnetic group**

_____   _____

_____   _____

_____   _____

_____   _____

_____   _____

_____   _____

_____   _____

Share your groups with the class.

Metals and magnets

Activity 1

You will need: a pen or pencil.

Complete these sentences using these words (you may need to use some words more than once):

> force   magnetic   wires   poles   metals   pull
> current   gap   glass   light   push   lamp
> flow   copper   like   switches   wood
> connected   plastic   paper   repel   magnet
> break   attract   electrical   opposites

1. Not all _____ are magnetic. _____ and gold are examples of metals which are non-_____.

2. You can use a _____ to find out if a material is magnetic. Most are non-magnetic, for example _____, _____ and _____.

3. Two _____ of a magnet will _____ one another if they are _____ and _____ one another if they are _____.

4. Magnetism is a _____ which can _____ or _____ objects.

5. To make an _____ circuit the items must all be _____ to one another without a _____ or _____ in the circuit.

6. We use _____ to start and stop the _____ of electricity round the circuit.

7. Some materials, such as glass and _____, will stop the _____ of _____ through a circuit.

8. It is important to connect the _____ to the _____ in the correct places to make it _____ up.

Chapter 5: Electricity and magnetism

Activity J

You will need: a pen or pencil.

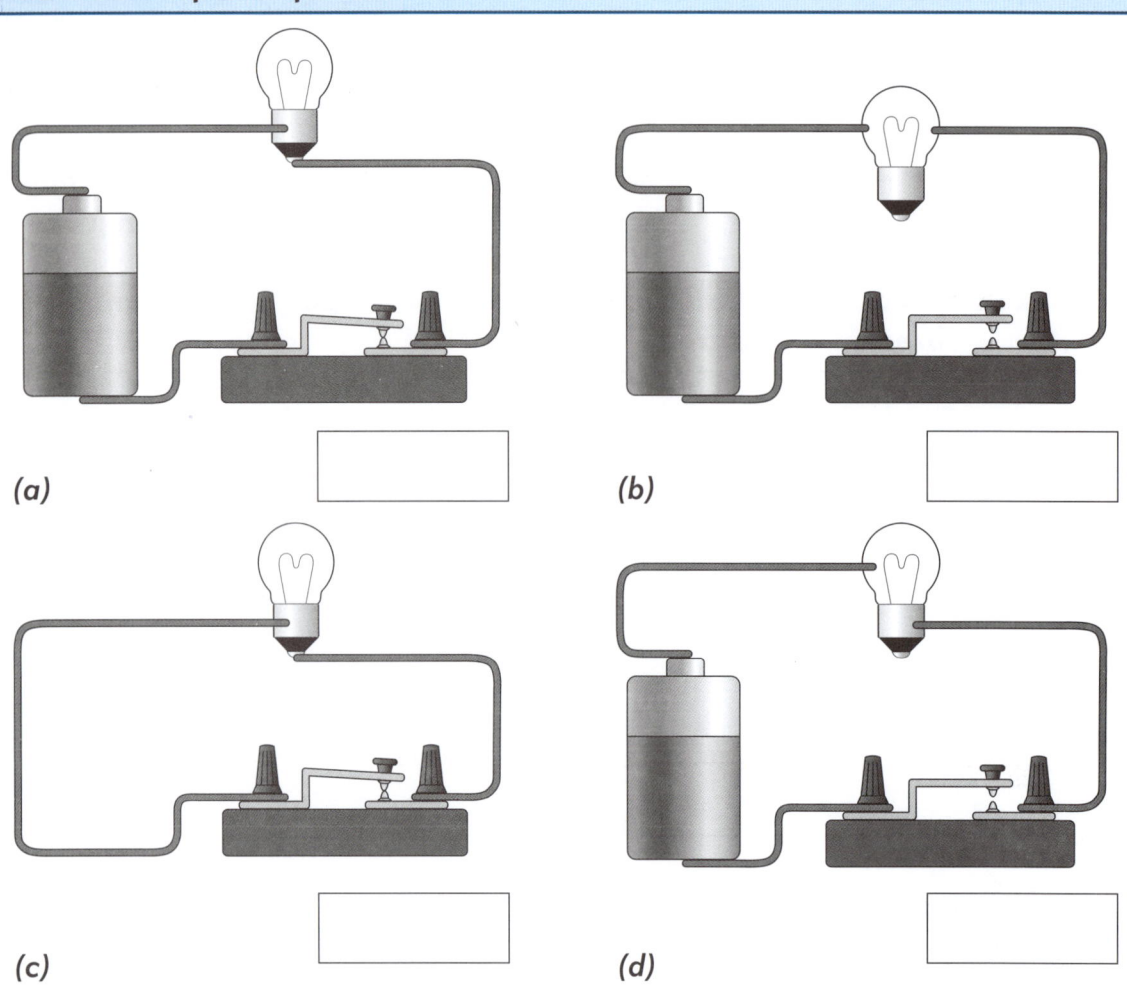

(a)   (b)

(c)   (d)

1. Look at the drawings of the four circuits. Which ones will light the lamp?

2. Write 'Yes' or 'No' in the boxes under the drawings. Think about why some lamps will not light.

3. Write a sentence about each circuit that will not light the lamp, to explain why it will not light. Use the space below for your answers.

_____

_____

_____

_____

_____